BEASTLY BATTLES

OSTRICH vs. VELOCIRAPTOR

Published by Capstone Press, an imprint of Capstone.
1710 Roe Crest Drive, North Mankato, Minnesota 56003
capstonepub.com

Copyright © 2025 by Capstone. All rights reserved. No part of this publication may be reproduced in whole or in part, or stored in a retrieval system, or transmitted in any form or by any means, electronic, mechanical, photocopying, recording, or otherwise, without written permission of the publisher.

Library of Congress Cataloging-in-Publication Data
is available on the Library of Congress website.
ISBN: 9781669088738 (hardcover)
ISBN: 9781669088998 (paperback)
ISBN: 9781669088776 (ebook PDF)

Summary:
Get ready for an epic battle! The ostrich faces off against Velociraptor. One stands as the largest bird living today. The other is a ferocious dinosaur that lived millions of years ago. It's a fight between epic beasts! Who will be victorious?

Editorial Credits:
Editor: Peter Mavrikis; Designer: Bobbie Nuytten; Media Researcher: Svetlana Zhurkin; Production Specialist: Whitney Schaefer

Image Credits:
Capstone: Jon Hughes, cover (bottom), 9, 13, 16, 17, 21, 27 (top); Dreamstime: Elizabeth Lombard, cover (top), Linda Bucklin, 28; Getty Images: janetteasche, 5 (top), Manoj Shah, 11, Stocktrek Images/Mark Stevenson, 5 (bottom); Shutterstock: Allexxandar, 19, Andrea Willmore, 27 (bottom), Daniel Eskridge, 25, Four Oaks, 29, Goldilock Project, 22, JamesFarleyPhotos, 7, Natalia Golovina, 10, Ryan M. Bolton, 18, Sergei25, 14, Sergej Razvodovskij, 15, Thomas Roell, 23

Any additional websites and resources referenced in this book are not maintained, authorized, or sponsored by Capstone. All product and company names are trademarks™ or registered® trademarks of their respective holders..

TABLE OF CONTENTS

MEET THE FIGHTERS	4
BIG BIRD	6
DANGEROUS DINO	8
RUNNING WILD	10
TOOTHY TERROR	12
KILLER KICK	14
TERRIBLE TOES	16
A GREAT VIEW	18
SHARP-EYED	20
STICK TOGETHER	22
SOLO AND SMART	24
FIGHT!	26
WHO'S THE WINNER?	28
GLOSSARY	30
READ MORE	31
INTERNET SITES	31
INDEX	32
ABOUT THE AUTHOR	32

Words in **bold** are in the glossary.

MEET THE FIGHTERS

Look out! It's a battle of mighty beasts.

One is the largest bird living today.

The other was a deadly **predator** from long ago.

Which would win this epic battle?

BIG BIRD

The ostrich is the heaviest bird in the world. It can weigh up to 300 pounds (136 kilograms). It's so heavy that it cannot fly.

The ostrich lives on the **savanna** in Africa. It's surrounded by hungry predators like lions and leopards. The ostrich has to be tough to survive!

DANGEROUS DINO

Velociraptor was a dinosaur. It went **extinct** about 70 million years ago. Velociraptor lived in the deserts of central Asia.

Velociraptor had feathers like a bird. However, this deadly dino was too heavy to fly. Instead, it ran on its **hind** legs.

RUNNING WILD

The ostrich is the fastest animal on two legs. It can run up to 35 miles (56 kilometers) per hour. This giant bird can cover up to 16 feet (4.9 meters) in one **stride**!

The ostrich uses its wings to help **balance** while running. This lets the bird turn quickly and escape any hungry beasts.

TOOTHY TERROR

Velociraptor was a dangerous predator. It fed on smaller animals and eggs. Velociraptor had a mouth full of razor-sharp teeth.

Velociraptor ran on two legs. It could run up to 24 miles (39 km) per hour. Few dinos could outrun a hungry Velociraptor.

KILLER KICK

The ostrich has long, powerful legs. And they can pack a punch! This bird will kick any predators that get too close. An ostrich kick is so strong it can kill a lion.

An ostrich has two large toes on each foot. One toe has a huge claw. This can be used to fight off any **threats**.

TERRIBLE TOES

Velociraptor had three long toes on each foot. One toe had a giant claw. It was more than 2.5 inches (6.4 centimeters) long. This claw helped it catch and hold **prey**.

This deadly hunter wasn't as big as some other dinos. It was about 6 feet (1.8 m) long and weighed 30 pounds (13.6 kg) or more. That's the size of a large turkey.

A GREAT VIEW

The ostrich is surrounded by hungry predators. It needs to be **alert** all the time. Good thing this big bird can see any threats that get too close.

The ostrich stands up to 9 feet (2.7 m) tall. It also has great vision. An ostrich eye is about five times larger than a human's eye! Few animals can sneak up on an ostrich.

SHARP-EYED

Velociraptors had great vision too. They had front-facing eyes. This gave them **binocular** vision. Binocular vision helped them chase down and capture prey.

Velociraptors also had a strong sense of smell. They were good at sniffing out prey. There was no way to hide from a deadly Velociraptor!

STICK TOGETHER

The ostrich is a **social** animal. These big birds like to live in **herds** of 12 or more.

Living in a herd means safety in numbers. Someone in the herd is always on the lookout for hungry lions or leopards nearby.

SOLO AND SMART

Velociraptors were not social animals. They did not hunt in groups. Instead they hunted alone.

This dangerous dino had a large brain compared to the rest of its body. Because of this, scientists believe the Velociraptor was very smart . . . for a dinosaur.

FIGHT!

On the savanna, the two beasts meet.

Velociraptor is hungry. It snaps its jaws, showing off its terrible teeth.

But a herd of ostriches surround the lone predator. Each scrapes at the ground, ready to kick. They're not going to back down!

WHO'S THE WINNER?

Now you've met the two beasts.

Velociraptor is fast and deadly. But the ostrich won't back down.

Which would win a beastly battle?

	Ostrich	**Velociraptor**
RANGE	Africa	central Asia
HABITAT	savanna	desert
DIET	mostly plants	smaller animals and eggs
WEIGHT	up to 300 pounds (136 kg)	30 pounds (13.6 kg) or more
SPEED	35 mph (56 kph)	up to 24 mph (39 kph)
WEAPONS	• sharp claw on one toe • strong kick	• huge claw on one toe • many sharp teeth
STRATEGIES	• great vision • herd animal • fast runner	• great vision and sense of smell • solo hunter • fast runner to catch prey • smarts!

GLOSSARY

alert (uh-LURT)—paying attention to surroundings

balance (BA-lanss)—to remain steady and in control

binocular (bi-NOCK-u-larr)—using both eyes with overlapping fields of view

extinct (ex-TINKT)—no longer living

herd (HERD)—a large group of animals that lives together

hind (HYND)—legs at the back of an animal's body

predator (PRED-uh-tur)—an animal that hunts other animals for food

prey (pray)—an animal that is hunted for food

savanna (suh-VAN-uh)—a large, grassy area with few trees

social (SOH-shuhl)—part of a group

stride (STRYD)—a long step while walking or running

threat (THRET)—possible danger or harm

READ MORE

Humphrey, Natalie. *Ostrich: Colossal Bird.* New York: PowerKids Press, 2023.

Jacobson, Bray. *Ostriches.* New York: Gareth Stevens Publishing, 2024.

Raatma, Lucia. *Discover the Velociraptor.* Ann Arbor, MI: Cherry Lake Publishing, 2023.

INTERNET SITES

African Wildlife Foundation: Ostrich
awf.org/wildlife-conservation/ostrich

National Geographic Kids: Velociraptor
kids.nationalgeographic.com/animals/prehistoric/facts/velociraptor

Natural History Museum (UK): Vicious Velociraptor
nhm.ac.uk/discover/velociraptor-facts.html

San Diego Zoo Wildlife Alliance: Ostrich
animals.sandiegozoo.org/animals/ostrich

INDEX

brains, 24, 29

claws, 15–16, 29

deserts, 8
dinosaurs, 8, 12, 24

extinction, 8

feathers, 8
flight, 6, 8
food, 12, 29

habitat, 6, 8, 29
herds, 22–23, 26, 29
hunting, 17, 24, 29

kicking, 14, 26, 29

legs, 8, 10, 12, 14

predators, 4, 6, 11–12, 14–15, 18–19, 23, 26
prey, 16, 20, 29

savannas, 6, 26
sense of smell, 20, 29
size, 4, 6, 8, 17, 19, 29
speed, 10, 12, 28, 29

teeth, 12, 26, 29
toes, 15–16

vision, 18–20, 29

wings, 11

ABOUT THE AUTHOR

Charles C. Hofer is a writer and biologist living in New Mexico, luckily far away from any Velociraptors.